Ever Blooming THE ART OF BONNIE HALL

Foreword by Robert Michael Pyle　　Edited by James D. Hall

OREGON STATE UNIVERSITY PRESS　　CORVALLIS

Ever Blooming

The Art of Bonnie Hall

The paper in this book meets the guidelines for permanence and durability of the Committee on Production Guidelines for Book Longevity of the Council on Library Resources and the minimum requirements of the American National Standard for Permanence of Paper for Printed Library Materials z39.48-1984.

Library of Congress Cataloging-in-Publication Data

Hall, Bonnie B. (Bonnie Birkemeier)

Ever blooming ; the art of Bonnie Hall / foreword by Robert Michael Pyle ; edited by James D. Hall.— 1st ed.

p. cm.

ISBN-13: 978-0-87071-116-9 (cloth : alk. paper)

ISBN-10: 0-87071-116-4

1. Hall, Bonnie B. (Bonnie Birkemeier) 2. Botanical illustration—Northwest, Pacific. 3. Wild flowers—Northwest, Pacific—Pictorial works. 4. Butterflies—Northwest, Pacific—Pictorial works. 5. Botanical artists—Oregon—Biography. I. Hall, James D. II. Title.

QK98.183.H34H34 2005

769.92—dc22

Printed in China

Text and jacket design by Erin Kirk New

Oregon State
UNIVERSITY

Oregon State University Press

500 KERR ADMINISTRATION

CORVALLIS OR 97331

541-737-3166 • FAX 541-737-3170

http://oregonstate.edu/dept/press

Contents

Scientific Names

Foreword

As the lucky husband of a botanical printmaker, I have had the bright privilege of getting to know a number of other women who practice the gentle art of depicting our Northwest flora and fauna through the medium of hand-wrought colors on paper. Several of these have shown their works at the Audubon Society of Portland's Wild Arts Festival, where I take part in the parallel authors' fair each Thanksgiving weekend. Thus I became acquainted with the exquisite silk-screens, and the effervescent person, of Bonnie Hall.

From the start, I was drawn to the clear, pure colors, emphatic forms, and what I can only call personality of Bonnie Hall's serigraphs. She has created one of the loveliest gardens ever grown of our region's floral treasures. From violets to shooting stars, fritillarias to ferns, and a bursting bouquet of irises, Bonnie captured the color, form, and essence of each of her subjects with the simplicity of a Henry Evans linocut and the exuberance of a Dutch Master's still life.

Another quality I especially appreciate in her work is the satisfying merger of science and art. Stemming from her professional background as a scientific illustrator, Bonnie's knowledge of the scientific details of her plant materials matched that of any botanist. But she also owned the eye of the true artist, and she somehow managed to imbue her flowers with both qualities to the betterment of each. As much as any artist I've known, Bonnie navigated with grace that high ridge where, as Nabokov wrote, "the mountainside of scientific knowledge meets the opposite slope of artistic imagination."

A few years after I met her, Bonnie acceded to lepidopterist John Hinchliff's request to illustrate the cover of his *Atlas of Oregon Butterflies*, and later the companion volume for Washington, with which I had been much involved. The Oregon atlas bears the official state insect, the Oregon swallowtail, which had earlier appeared in full color on T-shirts printed for the Oregon Entomological Society. On the jacket of the Washington atlas you will see the spectacular, cherry-spotted Clodius parnassian. Later she drew and printed two endangered butterflies, the Oregon silverspot and Fender's blue, as well as the beloved monarch. I was especially impressed that Bonnie managed to depict butterflies with all the dexterity and beauty that she brought to flowers.

As a beautiful bonus, in these pages, several of the splendid portraits of Northwest butterflies linger among the wildflowers they love. You will find the parnassians' bleeding hearts, the silverspots' blue violets, and a marvelous pairing of species named for two grand old Northwest naturalists: Fender's blue and its required host plant, Kincaid's lupine. In fact it was the gorgeous print of the violets that I chose to hang in my own study at home. The green and grapy impression exactly catches the shade, the softness, and the singular curl of the petals, the flying hearts of leaves and stems. I can almost smell them; can nearly feel the fritillaries' wingbeats as they alight on the violets they crave. Nabokov also wrote that "the highest enjoyment of timelessness . . . is when I stand among rare butterflies and their food plants. This is ecstasy." I know that ecstasy, and I find it here, among Bonnie's flowers and insects.

As one who attempts to capture some sense of living organisms through words, I always stand in awe of others who have the powers to depict them honestly and movingly in images. From the glyphs and runes of the ancients, to the etchings of Dürer and Bewick, to the oils of Van Gogh and O'Keefe, to the acrylics of

many contemporary workers, the tradition of biologically inspired art is long and deep. Bonnie Hall and her Northwest colleagues fit right into this honored guild of artists who would bring their love of the land and its plants and animals to bloom in pictures. Silkscreen is an especially demanding form because it requires its practitioners to forge exquisite detail from individual color fields, and Bonnie Hall succeeded in this exceptionally well. I don't see how anyone, versed in floristics or not, could look away from her flowers easily. Indeed, her booth at Wild Arts was often thronged with admirers reluctant to leave.

When I page through *Ever Blooming*, I feel as if I am rambling afield at the height of the blossom. My eye catches on Brown's peony, and I am immediately transported to Hat Point above Hell's Canyon near Imnaha. Its otter-brown and pinot noir petals carry me, through Bonnie's magic, deep into the evolutionary history of our own red garden peonies. The camas print takes us into meadows and swales running blue in May with those lilies' pyramidal inflorescences. The cobra lily and the golden iris, sharing sunny colors along with location, emplace us in the wild Siskiyous. Grass widows evoke spring trysts with my own botanist as the Columbia Gorge's wildflower eruption begins in March; lady's slipper, weekends at the foot of Mount Adams; larkspur and Columbia lilies, hot days afield among those night-blue and sun-gold fancies hung with two-tailed tiger swallowtails. Linnaeus himself would be thrilled by Bonnie's depiction of twinflower, the plant that charmed him in Lapland and bears his name: *Linnaea borealis*. Nor is this entirely a catalogue of bright colors from all over the floral spectrum: a quin-tych of ferns returns us to the calm and quiet of the forest, where—surprise!—we find an elegant lacebug clinging to the moss in the shaded understory.

I cannot omit to mention the delightful and well-written texts that accompany each print. Composed by Bonnie's hand, they show that her interest in her subjects ran far beyond mere form and color, into their natural history, human history, and etymology. These extended captions serve as a memorable mini-introduction to the Northwest flora and plant exploration. I also wish to note the careful and loving shepherding of these works by Bonnie's husband and partner in exploration, Jim Hall. An articulate biologist himself, Jim wrote the memorable preface and profile of the artist. Just as he always assisted Bonnie in exhibiting her work, Jim has ensured that her artistry will go on, through this lovely book. *Ever Blooming* is Bonnie and Jim Hall's great gift to all of us who love the Pacific Northwest and its wildflowers—long may they blossom!

ROBERT MICHAEL PYLE

"Some people, like flowers, give pleasure just by being"

—ANONYMOUS

Preface

Bonnie Hall was born an artist. From her earliest school days she showed an innate talent for capturing three-dimensional life and transforming it to the two-dimensional medium of paper. Through a nearly forty-year career as a scientific illustrator, she drew intensely detailed black-and-white drawings. Essential to science, they were published in scientific journals "very badly needed by very few people," as Bonnie said. When she discovered color serigraphy in 1991, she found her life's calling. Her screenprints of native wildflowers and butterflies were immensely popular. She was eager that a larger public come to appreciate the natural beauty of our Pacific Northwest landscape.

Bonnie died of cancer, too soon, at seventy-two. She was at the height of her artistic career and was just beginning to feel comfortable with the difficult art of screenprinting. Of the thirty-two large prints she produced, fourteen are now sold out, and others are nearly gone. This book is dedicated to her life and talent, in the hope that many more people will come to appreciate the natural world through her eyes.

BONNIE HALL—A BIOGRAPHIC SKETCH

Bonnie Carolyn Birkemeier was born November 18, 1931 in Portland, Oregon, to Edwin and Alice Tracy Birkemeier. She grew up in Milwaukie, Oregon, and had fond memories of time spent as a child on her grandparents' farm nearby. There her love for the natural world was first made manifest. In 1949 she graduated from Milwaukie High School, where she was editor of the school yearbook. She received a Bachelor's degree in biology from the University of Oregon in 1953 and a Master's degree in zoology from the University of California, Berkeley, in 1956. In her senior year at the University of Oregon she was editor of *Oregana*, the college yearbook, and was chosen for membership in Mortar Board, the senior women's honorary.

I met Bonnie in fisheries and aquatic entomology classes at Berkeley in 1953. We were married on September 25, 1955, in her church in Milwaukie. After my two-year stint in the Navy, we moved to Ann Arbor, Michigan, where I entered graduate school at the University of Michigan. She had developed her self-taught skills in scientific illustration while completing her Master's thesis on the life history of a tiny aquatic fly (Fig. 1). On the strength of that work, she was hired as a full-time illustrator in the University of Michigan Museum. She illustrated specimens for faculty in several divisions of the museum.

In 1963 I was offered a job on the faculty of the Department of Fisheries and Wildlife at Oregon State University, and we moved to Corvallis, along with Carolyn and Kate, our two daughters. From that point onward the family embarked on many outdoor adventures. They began with a week-long backpacking trip to the Wallowa mountains when the girls were five and seven years old. The outdoor world was always an important part of our life.

Our experience as an American Field Service host family in 1976–1977 rewarded us with the gift of another valued daughter and sister, Maria Rimini from Italy. Maria opened the door to the wider world, sparking our international adventures. There were sabbatical years in British Columbia, New Zealand, and Scotland and numerous shorter trips abroad, including one to witness Maria's wedding in Italy. These adventures continued right to the end of Bonnie's life.

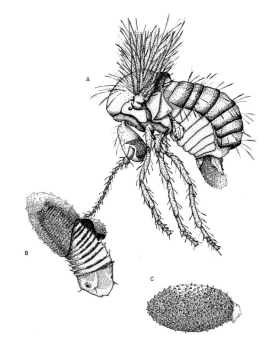

FIG. 1

Bonnie worked for thirty years as a scientific illustrator in the Department of Entomology at Oregon State University, where she produced many detailed black-line drawings of insects in support of research by faculty in the department (Figs. 2, 3). Coincidentally, much of her drawing was done for Professor Jack Lattin, who as a graduate student at Berkeley had been our teaching assistant in the entomology class in which we met.

In 1992, after a second bout with breast cancer, her artistic talent took a new turn, to screenprinting native Northwest wildflowers. This new phase in her professional life was stimulated by difficulties in screenprinting family Christmas cards in the 1960s and by a single disappointment in a later screenprinting venture. Bonnie was asked to design a print of the Oregon Swallowtail butterfly for the centennial of the Department of Entomology. She designed the stencils, but did not have the technical skill to complete the screenprint. This frustration provided the impetus for her to enroll in a one-term course in screenprinting at Linn-Benton Community College in the spring of 1991, two years before she retired from her work in the Department of Entomology. Sandy Zimmer

was an enthusiastic teacher and a great inspiration to Bonnie. During the course Bonnie studied the work of many print artists, including famed linocut artist Henry Evans, whose portrayal of wildflowers she truly admired. Among the screenprint artists she respected and learned from were Charley Harper, author of *Beguiled*

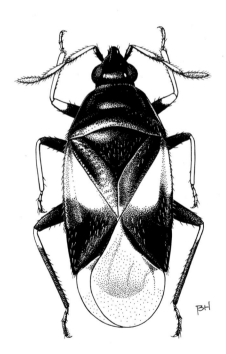

FIG. 2

by the Wild, Elton Bennett, and Oregon artists Sue Allen, Ellen Samms Burtner, Jim Howland, Donna Jepson-Minyard, and Earl Newman, who continues to produce screenprinted posters for the Monterey Jazz Festival and the Ashland Shakespeare Festival. Charley Harper provided her with some generous advice about technique in an exchange of letters.

Bonnie created her first serigraph, "Flags" while undergoing chemotherapy for her second breast cancer in the spring of 1992. (A serigraph is an original silkscreen print designed and hand pulled by the artist.) This was a deep purple rendition of *Iris tenax*, sketched from plants growing in our backyard. "My surgeon's advice for coping during chemotherapy was 'Don't do anything you don't want to do'—prompting my single-minded devotion to celebrating nature through screenprinting."

Bonnie created five botanical prints in her first year of printing. That November she invited friends to our home to see her first print renditions of native wildflowers. When they expressed a desire to buy her prints, Bonnie began her professional life as a screenprinter. She described the motivation and philosophy of her work in an artist's statement:

> A native Oregonian, I grew up charmed by the wildflowers of the region. There followed academic degrees in biology and a nearly forty-year career as a scientific illustrator. I retired with a will to share the privileged close scrutiny of nature that I had enjoyed.
>
> In embracing screenprinting, I have discovered the means. I love the construction of screenprinting, building an image one layer of

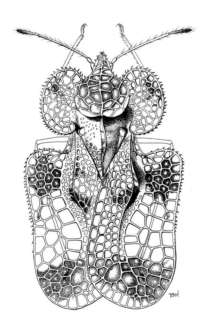

FIG. 3

color at a time. And I love the communication of it, creating multiple originals to entice a broader audience.

My subjects are native wildflowers and the occasional butterfly. In the tradition of natural-history portraiture, each rendition is faithful to form, color, and defining characteristics of the species. The ideal is an artful composition, rather like a family portrait that reveals personality, life stages, and the very essence of the subject.

Motivation for my work stems from a desire to celebrate the overlooked, undervalued, or threatened little natural treasures in our Pacific Northwest landscape. May it both delight and inform.

Bonnie showed her work at fine-arts fairs around the region and reveled in the conversations and connections that resulted. She would often entice prospective customers by telling them "these flowers are ever blooming," the inspiration for the book's title. Her prints were shown in several galleries and natural history museums. Five of them reside in the Northwest Art Collection of the Oregon State University Valley Library.

Bonnie remained a true scientific illustrator of each of her screenprint subjects and named her business "Scientific Serigraphs." The progression from observation of a plant to finished print was a lengthy one. She first sketched each plant in the field to capture its natural form. At this point came the crucial decision—the number of colors necessary to convey the essence of the plant or insect (each color requiring a different stencil in the printing process). Bonnie said that she often thought in color separation, a talent that enhanced her skill as a screenprinter. The butterflies were sketched from specimens in the Department of Entomology collection. She used a Pantone® color guide to ensure accuracy— though often, if asked about the color of one of her flowers, she would say, with a twinkle in her eye, "Somewhere out there is a flower that's just this color."

When her field sketch was completed, she visited the Oregon State University Herbarium to check on the details of the plant and to research its history. There and in the OSU Library were the sources used to develop the texts that accompany each print, texts that, in her words "give me additional pleasure in the researching."

Bonnie's prints portray the subjects with scientific detail, accuracy, and simplicity. A fellow artist remarked that she was able to capture the "gesture" of the plant. This critique pleased her, because she worked diligently to render her subjects as closely as possible to how they appear in nature.

Bonnie was a member of several scientific and arts organizations during her long career, including the Guild of Natural Science Illustrators, the Native Plant Society of Oregon, and the Corvallis Art Guild. In 1998 she joined the American Society of Botanical Artists, which provided some extraordinary opportunities and some frustration. A week-long trip to London with eight other ASBA members gave Bonnie the opportunity to see revered botanical prints in the Rare Print Room of the Victoria and Albert Museum, visit Kew Gardens and the Chelsea Physic Garden, and to meet with Shirley Sherwood, the world's foremost collector of contemporary botanical art. This was a deeply inspiring trip for Bonnie, one that opened her eyes to a whole new world. After the trip she developed an illustrated lecture on the history of botanical art that she presented to many audiences around the region. "Botanical Art: Three thousand years and one week in London."

But the Society of Botanical Artists was also the source of some distress. In November 1999 the ASBA published a proposed definition of botanical art. The last sentence read: "Silk screen works are *not* considered a medium in which it is *possible* to render botanical art" [emphasis added]. By that time a passionate advocate of screenprinting, Bonnie could not have disagreed more. She wrote a letter to the president that appeared in the next ASBA newsletter:

Dear Michele,

As a devoted screenprinter, I must react to the exclusive definition expressed in the November "Artist Update." Botanical illustration is about communicating as much information as possible in an artful manner. To single out any medium for exclusion means our options are limited to traditions of the past. For an organization that expects to make it into the twenty-first century, there must also be some room for innovation in the future.

I have been a scientific illustrator for nearly fifty years, albeit mostly entomological. I know that a good illustration is always a careful interpretation of reality filtered through a sensitive artist in league with whatever medium works. It is certainly valid to exclude specific media for a given exhibition. But to categorically exclude any one medium in an official definition of "botanical art" strikes me as both arrogant and myopic.

In defense of the targeted "silk screen works," is there a thorough understanding of what goes into a planographic stencil print? From field sketches to composition to interpreting the image as distinct defined colors, each translated into a stencil and hand-pulled separately with painstaking registration, it is by no means quick or easy. On the ASBA tour last spring I had the privilege of scrutinizing Ehret's original drawing of the Turk's-cap lily in the Rare Print Room of the Victoria and Albert Museum. His delineation of overlapping colors is not unlike a silkscreen print. I like to think that he would have relished the medium. Imagine being able to personally create multiple originals, retaining control over the coloration of each. And what about Henry Evan's linoleum-block prints? No one has conveyed grasses more succinctly. Some believe that a minimalist rendition may actually give the viewer a clearer scientific picture of the defining characteristics of a specimen.

I urge you not to adopt an exclusive definition for what will be considered acceptable media for botanical art.

The editor of *The Botanical Artist*, newsletter of the ASBA, was also persuaded, putting Bonnie's print of the Cobra lily on the cover of the issue in which her letter appeared. Others, including the president, wrote in a similar vein. There seemed to be little support for the provisional definition. The ASBA editor, an instructor in the Certificate Program in Botanical Art and Illustration at the Denver Botanic Gardens recently remarked that Bonnie's letter and further phone and e-mail conversations had supported her efforts to teach botanical art as a thriving, changing, and contemporary genre, encompassing many media.

A couple of years later, still feeling that screenprinting was a poor cousin in the ASBA, Bonnie prepared a draft of an article she planned to submit to the newsletter, but it was never sent:

Printmakers in Our Past (and Present?)
September 20, 2002

*"In the digital age of scanners and color copiers, the
printmaking process can barely be appreciated."*

Three events prompted the following submission:

1. The above quote from a *Denver Post* review
of an exhibit of plates from "Temple of Flora"
currently at the Denver Botanic Gardens.

2. Reading I did for a talk about printmaking
in the history of botanical illustration given at
the University of Oregon Museum of Natural
History.

3. The haunting memory of a proposed
Fall 2000 ASBA exhibit of botanical prints in
Connecticut that was cancelled even before the
submission deadline because applicants confused
"prints" with photo-mechanical reproductions.

Admittedly, common usage is at fault. The term
"print" is applied equally to the labor-intensive,
hands-on, multiple original, as well as to facsimile
reproductions of a unique original piece.

Like no other art form, botanical illustration
has a rich legacy of printmakers. The earliest
printed herbals included woodblocks of medicinal
plants. Albrecht Dürer brought the Renaissance to
northern Europe with engravings of naturalistic
images incised in wood and metal. The famous
pasque flower attributed to his student Weiditz
was created by cutting away all the wood except
for those delicate black lines. It was art with
a purpose—to increase distribution through
multiple images, communicating information
about plants in an artful manner. Some botanical
artists did not do their own printmaking, but
collaborated with engravers in a synthesis of arts.

Many of our heroes did do their own
printmaking. Emanuel Sweerts (my alter ego
because he also sold his prints at street fairs) in 1612
published one of the first florilegia, a collection
of engravings of uneven quality done by himself
and others. Maria Sibylla Merian, daughter of
an engraver and stepdaughter of a famous Dutch
flower painter, documented her fascination with
insects and their host plants by engraving the
plates for a volume on *The Wonderful Transformation of
Caterpillars*. Elizabeth Blackwell, striving to redeem
her husband from debtors' prison, produced
A Curious Herbal, five hundred plates of plants
drawn at the Chelsea Physic Garden. She
personally engraved, printed, and hand-colored

each copy and paid off her husband's debts, whereupon the ingrate left her. Then came the Golden Age of plant illustration, when scientific botany and printing quality and artistic talent aligned. Georg Dionysus Ehret, unequaled for accuracy and excellence, and Pierre Joseph Redouté, perhaps the most famous name in botanical illustration, both did some of their own engraving. Redouté espoused a stipple technique that produced a softer and more luminous print. In the nineteenth century, that most prolific Walter Hood Fitch, Glasgow calico designer turned botanical lithographer, produced some ten thousand illustrations, many for *Curtis's Botanical Magazine*, drawing with grease crayon directly on the lithographic stone. And then there is my personal hero, Henry Evans (1918–1990), who took the elementary medium of linoleum-block printing to exquisite heights.

Are there other botanical printmakers out there somewhere today? As enumerated in a 1986 Hunt Institute publication *Printmaking in the Service of Botany*, they might be printing in *relief* (woodblock or linoleum), *intaglio* (engraving, etching, mezzotint, etc.), *planographic* (lithograph or screenprint), or *nature printing*. I am a screenprinter feeling rather lonely in the ASBA.

Bonnie became an ardent member and supporter of the Native Plant Society of Oregon, and it was a wonderfully symbiotic relationship. Society members helped in her quest for plant species to illustrate, and she returned the favor. She contributed prints that were used in fundraising for the Oregon Flora Project and provided designs to support fundraising for local chapter and state activities. Many society members became her good friends.

In a tribute article published in the *Corvallis Gazette-Times* after her death, correspondent and friend John Ginn noted that many local artists may not have realized how much Bonnie had done to promote the artistic community. She often pointed him to interesting stories about local artists. It was this support, mentoring, and promotion, among other activities, that led her to be named Corvallis Patron of the Arts for 2003 at a community celebration in January 2004. She was especially pleased with this award, recognizing as it did the many ways in which she had contributed to the arts.

Other honors have followed. The OSU Department of Botany and Plant Pathology, in recognition of her generous donation of prints and note cards to be sold for the benefit of students, has named the resulting endowment fund the Bonnie Hall Student Activity Fund. The Corvallis Fall Festival, in which she participated as an artist for many years and more recently as a member of the Board of Directors, presented the first annual *Bonnie Hall Best in Show* award in September 2004. Six of her prints will be enameled in fused glass and made a permanent part of the Madison Avenue Alley Art, a project of the Corvallis Arts Center.

Bonnie was generous with her artwork, donating a portion of the proceeds from particular prints to many conservation organizations. Among these were the Nature Conservancy, Xerces Society, Native Plant Society of Oregon, Siskiyou Project, and the Greenbelt Land Trust.

Bonnie was diagnosed with pancreatic cancer in December 2003 and died peacefully in her beloved home on February 18, 2004. The life she led was true to her love of nature and her desire to conserve its beauty. Her message was communicated by example, more clearly than had she preached it widely. Bonnie left the wider community a beautiful legacy of wildflower prints to remind us of our rich natural heritage. And she left family and friends the legacy of a life well and fully lived.

JIM HALL
JUNE 2005

Ever Blooming THE ART OF BONNIE HALL

Brown's Peony, Western Peony *Paeonia brownii*

There really is a wild peony in the Pacific Northwest, but only just this one species. It is big and showy, and yet easily overlooked. No wonder the intrepid Scottish plant collector David Douglas considered finding the peony one of the more important events of his travels. Just twenty years after Lewis and Clark wintered on the Pacific Coast, Douglas systematically explored some of the same territory alone, gathering specimens for the Horticulture Society of London. He collected the peony in 1826 in the Blue Mountains and honored it appropriately with the name of Robert Brown, a fellow Scot and eminent British botanist of the time.

I first met Brown's peony along the Little Blitzen River on Steens Mountain, where striking dried seedpods flanked the trail in late summer. To see foliage and flowers required other places in other years. Look for the deeply incised thick green foliage sheltered under sagebrush and pines on the drier eastern side of the Cascade Range and throughout much of the arid Far West. Flowering is in May and June. In bud, flower,

and mature fruit the heavy heads droop to the ground, making them inconspicuous.

This herbaceous perennial forms multistemmed clumps up to two feet tall, each fleshy stem graced with a single flower. Exceptions were seen in larger, presumably older clumps where an additional flower stem originated at the base of some lateral leaves. The robust flowers tend not to open fully. Tight concentric rings of purple-tinged sepals and yellow-rimmed maroon petals encircle five green pistils and numerous golden stamens. Mature fruits are clusters of five to six leathery follicles containing large blackish seeds. The unusual appearance of Brown's peony, a close cousin of the well-known ornamental garden plant, makes it easy to identify.

Brown's Peony   2003

California Poppy, Copa de Oro   *Eschscholzia californica*

Here is perhaps the most celebrated West Coast wildflower. Called Copa de Oro by the earliest Hispanic inhabitants, this golden poppy once emblazoned such vast fields that its brilliant color was visible from sailing ships miles offshore—a plausible basis for the early name for California, La Tierra del Fuego, land of fire. Discouraged by grazing, agriculture, and development, it now blooms in lesser abundance, May through September, from Southern California north to the Columbia River, and elsewhere where it has escaped from garden plantings.

This marvelous perennial arises from a deep taproot and generally reaches two feet in height. The flowers are two to three inches across and vary from pale to deep yellow and orange. They are remarkably responsive to sunlight, closing at nightfall or in overcast weather. Sepals are united in a conical structure that is thrust off by the opening petals, much like doffing a little hat. The seed capsule is linear and tipped by the withering styles. The foliage is grayish-green and finely divided.

Eschscholzia californica owes its introduction to the botanical world to three explorer naturalists and a coincident Russian. Archibald Menzies, Scottish botanist and surgeon with Captain George Vancouver, was first to collect specimens for transport to foreign shores, in Monterey in December of 1792. But his classification was incorrect, his collection suffered badly on shipboard, and the plants delivered to Kew Gardens soon died. Next was Adelbert von Chamisso, French naturalist with the Russian Romanoff expedition, who spent October of 1816 at San Francisco. Here he collected, described, classified, and named *E. californica* after the ship's noble young surgeon, Johann Friedrich Eschscholtz. Finally, it came to David Douglas to convey this vibrant poppy to the world. Collecting along the Multnomah [Willamette] River in 1825, he first encountered *E. californica* and sent back to the Horticultural Society of London the seeds that have thrived in English gardens and beyond.

Copa de Oro   1992

Calypso, Fairy Slipper Orchid   *Calypso bulbosa*

A deceptive beauty, this tiny orchid provides no reward for those insect pollinators that would ensure its reproductive success. Though the false nectar spurs and shooting star-like blossom promise sustenance, no nectar is produced. Only first-time visitors (probably inexperienced bumblebees) can be fooled. Other orchid gamblers also ply this deception, relying on the huge numbers of seeds that result from an occasional pollination, while avoiding the costly production of food rewards.

The only species in the genus, *Calypso bulbosa* occurs in north temperate regions the world around. It was first described by Swedish botanist Carl Linnaeus and named after the sea nymph of Homer's *Odyssey*, inspired by a propensity for seclusion shared by orchid and Greek goddess. In the Northwest, *C. bulbosa* ranges from Alaska into California, in the deep shade of cool, moist forests, from near sea level to midmontane elevations. In the fall, this exquisite perennial produces a single broadly ovate leaf from a bulbous corm. In March or April, solitary magenta blossoms appear, delicately fragrant, the slipper-like lip prolonged below into two short spurs. Flower stems range in height up to eight inches. For the persistent observer the unusual white form is a thrilling discovery.

The flower's very attractiveness threatens extermination. The species is considered rare, listed by Britain's Kew Gardens as vulnerable to extinction on a global scale. Locally there are some precious areas of abundance. Transplanting is not successful, even though the plant is easily dislodged, as it sits lightly on the carpet of leaf mold and moss, with roots scarcely penetrating. Cultivation efforts fail because the link is broken between the orchid's mycorrhizal roots and specific sustaining fungi in the substrate. Hands off!

Calypso 1994

Checker Lily, Rice-root Lily, Mission Bells   *Fritillaria affinis*

If one set out to design a flower, chartreuse and brownish purple would not likely be the colors of choice. In this unusual lily, however, the effect is quite beautiful and ultimately practical. These colors contribute to a ruse. In the intimate experience of sketching the plant in a roadside ditch, I became aware that the flowers exude a clear syrup that smells very bad. Flies were attracted and, bumbling into the blossoms, they must serve as pollinators. Foul smell and putrid colors conspire to ensure perpetuation. Whatever works!

It's the plant's names that are descriptive. "Mission Bells" envisions the pendant bell-like flowers and their distribution from California's mission country in the south, west of the Cascade Mountains in Oregon to southern British Columbia, and eastward in Washington, northern Idaho, and Montana. The geometric pattern surely inspired "Checker Lily" as well

as the scientific name *Fritillaria,* from the Latin *fritillus* (dice box), and "Rice-root" refers to the tiny bulblets resembling grains of rice that cluster around the conical white bulb.

These little rice grains give rise to a colony of leaflets that accompany the tall flower stalk—that is if a hungry deer has not lopped it off. The elegant seed pods are papery, broadly winged vessels containing six ranks of flat seeds. Found in prairies, woodlands, and coniferous forests from sea level to 5,000 feet elevation, this is a species fast disappearing throughout its range. Not until you see its incredible abundance in a surviving fragment of undisturbed native habitat can you imagine what it must have been like when Meriwether Lewis first collected it on Brant [Bradford] Island in the Columbia River.

Fritillary   1996

Clodius Parnassian Butterfly, Bleeding Heart   *Parnassius clodius, Dicentra formosa*

Clodius and its host make one of the great aesthetic partnerships of the Northwest woods: the elegant green stands of lacy bleeding heart, flecked with aromatic pink flowers, frequented by the great, floppy white butterflies, speckled with scarlet and jet.

—Robert Michael Pyle, *Wintergreen*

These eloquent words by an impressive lepidopterist provided the inspiration for this piece. It is a portrait of a member of the swallowtail family that has managed to thrive in a most unfriendly habitat for butterflies. With up to a three-inch wingspan and leisurely flight, the Clodius Parnassian is conspicuous in the cold wet forests of Pacific Northwest mountains, ranging from Alaska to central California and east to western Montana and Wyoming.

Guided by some uniquely butterfly sense, adult females deposit eggs near clumps of wild bleeding heart, *Dicentra formosa* in this case. It is the essential larval food plant. Larvae are black and velvety with yellow spots on each segment, resembling a common cyanide-producing millipede. Benefiting from a case of mistaken identity, they are likely spared some predation by this ruse. The smooth brown pupae encased in loose cocoons may be found amid debris on the forest floor. Adults fly from May to late summer, and the cycle repeats.

It was the Russians, colonizing in Alaska and northern California, who were first to introduce this butterfly to science. Edouard Ménétriés, Conserver of Rarities at the Zoological Museum in the Imperial Academy of Sciences, St. Petersburg, classified the species in 1855 from specimens collected by I. G. Vosnesensky near San Francisco Bay. Parnassians are customarily named for Roman gods, but here Ménétriés commemorated a mere mortal—Clodius, a notorious rabble rouser in league with Caesar. Or might it have been his beautiful but scandalous sister Clodia? The edition title "Clodia" is a deliberate aberration justified by the fact that the specimen illustrated is, after all, a female.

Clodia   1994

Cobra Lily, California Pitcher Plant *Darlingtonia californica*

Exploring along a forest road near Cave Junction in southwest Oregon, we were startled by great colonies of these bizarre plants on hillside seeps. They appear in crowds of thigh-high green tubes, each spiraling upward into an ominous arched hood complete with ventral mouth and drooping bilobed moustache. Many generations persist, all packed together in a fantastic jumble. A passing shower gave voice to the assemblage, the murmur of raindrops drumming on dried hoods. Here were the creatures immortalized by Ken Kesey in *Sometimes a Great Notion* as "an artist's conception of chlorophyll beings from another planet . . . a creature trapped in that no-things land between plants and animals . . . sweet and sleek carnivore with roots . . ."

Darlingtonia californica is an insectivorous plant, acquiring some degree of nutrition from the capture of insects. Adaptation to this end is elaborate. The leaves have formed effective pitfall traps where insects are lured into the orifice by a profusion of nectar glands. Light entering through translucent spots in the hood beckons them on to a slick one-way descent enforced by downward-pointing hairs and ending with a fatal pool of water in the base of the tube. There bacterial decomposition renders the prey digestible. With blatant hypocrisy, a parallel enticement is offered to insects for the purpose of pollination. The stately solitary flowers arise in May and June, well endowed with scent glands. Their close-fitting petals are sculpted with opposing notches to form another welcoming entrance tunnel—this time benign.

Other members of the pitcher-plant family (Sarraceniaceae) occur in the eastern United States and Canada and in northern South America. Only this single genus and species occurs in the West, in southwest Oregon and northern California. No relatives span the intervening distance, suggesting that the plant developed millions of years ago, before its ancestors migrated westward. It ranges from the Sierra Nevada and Klamath Mountains to sea level, in swamps, bogs, and seeps with high-acid and low-nutrient soils, where annual rainfall exceeds forty inches. Threats come through destruction of habitat and wholesale collecting, in wanton disregard for this magnificent megaflora of such great antiquity.

Cobras   1996

Deltoid Balsamroot *Balsamorhiza deltoidea*

Spring at the east end of the Columbia River Gorge means hillsides painted in sunshine yellow. Tall clumps (two to three feet tall) of this perennial desert sunflower rise from thick taproots to festoon open areas and oak woods, blooming in April and May. *Balsamorhiza deltoidea* also occurs between Coast and Cascade ranges from southern Vancouver Island to southern California. The specimen from which Thomas Nuttall named and classified the plant in 1840 came from "near the outlet of the Wahlamet [Willamette]." Balsamroot was an important food plant for native tribes of the arid interior. They ate the tender young shoots and roasted the resinous roots and oily seeds.

The genus name *Balsamorhiza* is in reference to the sticky sap of the thick taproot, a veritable cable of vital filaments in a tough leathery casing. Leaves of this species are broadly triangular, nearly hairless, and with margins partly scalloped or with shallow rounded teeth. The sunflower head itself is really not one but a whole bouquet of flowers all precisely arranged in a tight nosegay. A close look reveals the spiral ranks of inconspicuous disk flowers. Lined up around the rim are the showy ray flowers, each flying a bright yellow "petal" or ray. Diverse elements united create such a sunflower and claim membership in the family Compositae.

While working to portray this radiant creature, I was blessed with its sunlight all during a month that proved to be the rainiest May ever recorded in Corvallis. Would it not be wonderful to have it brighten our home garden? Alas, according to Arthur Kruckeberg in his definitive *Gardening with Native Plants of the Pacific Northwest*, while that is successful in drier eastside gardens, those of us west of the Cascades are generally doomed to failure unless soils are dry, gravelly, and well drained. Further, "The whole plant takes poorly to transplanting, thus seeds should be used." I hope to try.

Balsamroot   1998

Douglas Iris   *Iris douglasiana*

Here is one tough little iris. Found in the far southwest corner of Oregon and south through central California near the coast, it flourishes on arid cutbanks, in heavily grazed pastures, and on grassy seaward slopes pummeled by gale-force winds. Cattle won't eat the bitter foliage but, according to one local farmer, they do like to lie down on the clumps.

Iris douglasiana is but one of eleven closely related species of Pacific Coast irises. Hybridization between contiguous species occurs in nature and there is an ongoing struggle to define distinct taxa. Douglas iris appears in highly variable guises. The showy flowers may range in color anywhere from cream through lavender to deep reddish purple. The leaves tend to be distinctly broader than those grass-like blades of the other species. Native irises make delightful and undemanding perennials for home garden use.

It is David Douglas whose name is given to this iris, the Douglas of the fir tree. This bold Scottish gardener, apprenticed to the leading horticulturists of his day, came to explore the wilds of the American West Coast in 1824 at the behest of the Horticultural Society of London, primarily to collect exotic plants for cultivation. Over a span of ten years he matched the challenges of wilderness, canoeing an unbridled Columbia River, befriending the native people, walking up to fifty miles in a day. At thirty-five, Douglas died on the island of Hawaii enroute home , but not before his prodigious collections had enriched the world's gardens and catalogued the natural history of a primeval Pacific Northwest. Common, sturdy in stature, thriving in harsh surroundings—these are all apt descriptions for the explorer naturalist and his namesake iris alike.

Douglas Iris 1995

Fender's Blue Butterfly *Icaricia icarioides fenderi*

The Blues are small jewels in the world-wide family Lycaenidae. *Icaricia icarioides fenderi*, Fender's Blue, named for the late naturalist Kenny Fender, is a race found only in a few small isolated colonies at the western edge of the Willamette Valley. It was first described in 1931 near McMinnville, Oregon, then thought to have become extinct, and has only recently been rediscovered. An exhaustive search has uncovered twelve locations and a total population estimated to be between three and five thousand individuals.

With wings spanning no more than an inch, this is one of the smallest butterflies. Only the males are blue above; females are uniformly light brown. Both have a tannish underside marked with black spots narrowly ringed in white.

One brood is produced each year and it is the second instar larva that overwinters. Further growth, pupation, and emergence occur the following year. Several species of ants associate with *I. icarioides*, feeding on secretions from the larval honey glands. The arrangement may be mutually beneficial, as ants have been known to attack potential parasites of the larvae. These smooth, slug-shaped caterpillars are restricted to various lupines as food plant—any given population of butterflies is usually confined to a single species of lupine. Our Fender's Blue is associated with Kincaid's lupine (*Lupinus sulphureus kincaidii*), which is found in the Willamette Valley native upland prairie, an ecosystem that is now reduced to less than one percent of its historic dimensions.

JH: In January 2000, this species, along with its host plant, Kincaid's lupine, was placed on the Endangered Species List by the U.S. Fish and Wildlife Service. The lupine was listed as threatened, the butterfly as endangered, the only butterfly in Oregon to be so designated (the only other butterfly in Oregon on the list is the threatened Oregon Silverspot, listed in July 1980).

Fender's Blue   1993

Giant Purple Trillium   *Trillium kurabayashii*

Like some botanical Bigfoot, "Giant Purple Trillium" conjures up the image of an elusive megaflower. And so it is. Search in moist woods of Curry County in the farthest southwest corner of Oregon and on the western slopes of the Klamath Mountains in adjacent California. A disjunct population may be found in the Sierra Nevada in Placer County. The reward is truly spectacular—deep purple petals nestled in a whorl of three great leaves mottled like assorted green pieces of a jigsaw puzzle, rising elegantly up to twenty inches on a sturdy bare stem. Flowering peaks in April.

A number of amazing variations on the trillium theme occur in temperate forests of the Northern Hemisphere in three unconnected areas: eastern Asia, western North America, and eastern North America. Seven or eight distinct species are known from each of the first two. By far the greatest variety (thirty-five species) can be found from the Midwest eastward. Trilliums are broadly divided into those plants with the flower held above the leaves on a thin stalk or pedicel, and those without pedicels (sessile).

In the case of this sessile giant, the Latin generic name *Trillium* is indisputable, the flower parts so characteristically in threes, but the species name continues in contention. Formerly lumped with *T. chloropetalum*, a new species named *T. kurabayashii* was designated for it in 1975. The name commemorates Japanese cytologist Masataka Kurabayashi, whose work on trillium chromosomes provided clues to its distinctiveness. In 1993, an all-new comprehensive volume, *The Jepson Manual, Higher Plants of California*, again lumped it, this time with *T. angustipetalum*. Called by whatever scientific name, the Giant Purple Trillium remains a natural treasure, a rare and stunningly beautiful creation commanding attention.

Giant Purple Trillium   1995

Golden Iris   *Iris innominata*

You can't judge an iris by its color. Gold is but one of the bright variations, from yellow to shades of violet, that adorn this species, but it is the golden form that festoons the wilderness trail along the Rogue River in May with a massive floral display. The dark green leaves are grass-like and graceful, the flowers large and with dark veins covering the falls. This is but one of a colorful handful of closely related species of native iris on the West Coast. Their propensity to hybridize has confounded taxonomists and delighted gardeners.

Iris innominata, literally the iris of no name, is forever linked with pioneer botanist and avid plant collector Lilla Leach. In the 1920s and 1930s she and her husband and two personable burros explored the wilds of the Siskiyou National Forest in the southwest corner of Oregon, an isolated area largely undiscovered by earlier naturalists. Following only scant trails in steep and rocky terrain, they collected and preserved plants, discovering this iris and other species that had never before been collected. It gives me a particular thrill to encounter the very specimens prepared by Lilla Leach nearly seventy years ago, now housed in the Oregon State University Herbarium.

The house and garden built by Lilla and John Leach overlooking a ravine on Johnson Creek in southeast Portland is today the city's first public botanical garden, a microcosm of the natural ecosystems they explored. Nine acres of trails and plantings provide a wilderness in the midst of the city. It is shepherded by the Leach Garden Friends, a group dedicated to public awareness of our unique Northwest flora. Leach Botanical Garden is a treasure appropriate to perpetuate the memory of a legend in the natural history of Oregon.

Golden Iris   1996

Grass Widows, Purple-Eyed Grass *Olsynium douglasii*

Dainty relatives of the iris, these plants grow in clumps less than a foot high, lost among the grasses until flowering blows their cover. My informal poll suggests that the popular definition of the term "grass widow" is not familiar to most people. The dictionary lists several definitions, among them a woman whose spouse is just away rather than deceased. There doesn't seem to be direct literal justification for so naming this plant, but it provides a pleasing visual metaphor. The scientific name also bears some explanation. The accustomed generic name *Sisyrinchium* has given way to *Olsynium*, making this the only representative of that genus to be found in the Northwest.

I was charmed by grass widows in the Columbia River Gorge where, beginning as early as late February, their massed blooms color the overlooking plateaus in luscious purple. Here in 1826 David Douglas first collected them "Near the Great Falls of the Columbia." That was

Celilo Falls, now submerged in the backwater behind The Dalles Dam. In addition to this incursion into the Gorge, they occur both west and east of the Cascade Range from Vancouver Island to Northern California in apparently dry places (prairies, rocky slopes, open oak and pine woodlands) but mostly where there is ample moisture in early spring.

According to Art Kruckeberg in his *Gardening with Native Plants of the Pacific Northwest*, this "exquisite plant" is a desirable candidate for the home garden. It is easily propagated by transplanting offshoots of the short rhizome. If further enticement is needed, consider that the satiny sheen of the delicate symmetrical blossoms glistens in sunlight, and the gentlest breeze sets them fairly dancing. With luck, a rare white-flowered clump may be encountered in the wild.

Grass Widows   2002

Green-banded Mariposa Lily *Calochortus macrocarpus*

Some sixty species of the genus *Calochortus* (literally "beautiful grass") range in western North America from Canada to Guatemala. *Calochortus macrocarpus* is widespread and abundant, found over much of the arid interior of the Pacific Northwest. It was introduced to botanists in 1828 through the heroic collecting of David Douglas, who first recorded it at the "Great Falls" (the Celilo Falls) of the Columbia River.

One hundred and sixty-seven years later, I was treated to a showing in late June along the lower Deschutes River. Here the Deschutes is a rambunctious torrent grinding its way north to the Columbia through great red-rock canyons rivaling those of the American Southwest. On a parched, sagebrush hillside above a bend in the river, overarched by a soaring red cliff opposite, these exquisite green-banded mariposa lilies were in fleeting bloom. It is hard to imagine a more delicate flower or a more hostile environment.

The plant is one to two feet tall. The petals are lavender and intricately patterned inside with highly ornamented basal glands, purple transverse stripes, and slender yellow hairs. In beauty of design and color they may have been likened to the wings of a butterfly, giving rise to the common name of *mariposa*, the Spanish word for butterfly. A precise longitudinal green stripe bisects the outside of each petal. Long, slender, three-angled seed capsules are held erect.

Attuned to harsh conditions, plants have even been found to produce more flowers per stalk following a fire. Indeed, they don't tolerate the least luxury. All attempts to transplant the deep-seated bulbs to home gardens have been unsuccessful. Called "sweet onion" by the Native Americans, the crisp, sweet bulbs were historically dug in spring before flowering and eaten raw or added as flavor in cooking. Unfortunately, this tall perennial is highly palatable to livestock, and heavy grazing and cultivation have reduced populations.

Mariposa   1995

Henderson's Shooting Star, Bird Bill   *Dodecatheon hendersonii*

Shooting stars are easy to spot. Bright petals swept back meteor-like from a dark beak define the plant, whether found nestled at the base of a weepy rock cliff in Zion National Park, massed in a magenta carpet beside a Cascades mountain lake, or accompanied by camas and buttercups in Corl's pasture, Corvallis.

The leaves of *Dodecatheon hendersonii* are somewhat fleshy and rounded, forming a tidy flat rosette. From this, a sturdy four- to twelve-inch flower stem arises, crowned with a cluster of two to fifteen reflexed blossoms borne on recurved stems. Flower parts are generally in fives, occasionally fours. The fruit is a one-celled capsule held erect and containing many seeds.

Pliny the Elder, in his thirty-seven-volume *Natural History* written in the first century A.D., reports the name *dodecatheon* from the Greek *dodeka* (twelve) and *theos* (god) "as a compliment to the grandeur of all the twelve gods." Linnaeus must also have found the plant worthy, for he officially assigned the name to this genus in the primrose family. Our particular species is named for pioneering Northwest botanist Louis F. Henderson.

D. hendersonii may be found March through June in open woods and prairies from Vancouver Island south on the west side of the Cascades to southern Oregon, and in both the Coast Range and Sierra Nevada to southern California. This exuberant little perennial makes a pleasing garden subject and is easily grown.

Shooting Stars   1993

Hookedspur Violet   *Viola adunca*

Sadly, the exquisite fragrance of this widespread little violet cannot be conveyed in printer's ink. It was bestowed lavishly on the illustrator. In mild years this perennial can blossom all winter, thus providing subject matter in a lean season. It has the greatest range of any native North American violet, occurring in Eastern Canada and New England and nearly everywhere in the West.

Viola adunca was first collected by one of the earliest of explorer naturalists, Archibald Menzies, who sailed to the Northwest with England's Captain George Vancouver aboard the *Discovery* in 1792. Specimens were brought to James Edward Smith in London for the initial description, published in Rees' *Cyclopaedia* of 1817:

> Stems simple, ascending. Leaves ovate, somewhat heart-shaped, obtuse, crenate, downy, dotted. Stipulas loosely fringed. Flower-stalks longer than the leaves. Nectary hooked. The two lateral petals are downy at the base.

The name *adunca* derives from *aduncus*, meaning "hooked," and refers to the spurred base of the lowermost petal. It is this species, where it occurs along the Oregon coast north of Florence, that is the preferred food plant for larvae of the threatened silverspot butterfly, *Speyeria zerene hippolyta*. Here the violet requires considerable direct sunlight and is excluded by encroaching brush. Conservation of *V. adunca* is essential to survival of the butterfly.

Blue Violet   1992

Kincaid's Lupine   *Lupinus sulphureus kincaidii*

Our native lupines are a confusing lot, with nearly six hundred distinct kinds described in the U.S. Bewildering variations occur within a given population, and rather free interbreeding compounds the confusion. One iridescent blue butterfly is not confused, however. Known to prefer Kincaid's lupine as a larval food plant, the Fender's Blue butterfly, *Icaricia icarioides fenderi*, was last seen in 1937 and presumed extinct. But careful scrutiny of populations of Kincaid's lupine led to rediscovery of the rare insect in 1989 by Oregon State University entomology professor Paul Hammond. Now efforts are underway to protect the butterfly and its host lupine.

This variety of the sulphur lupines is confined to remnant bits of native prairie grasslands in western Oregon and Washington, primarily in the Willamette Valley. Lupines are broadly recognized by radiating leaflets and blossoms that resemble those of other members of the pea family. Kincaid's lupine may be distinguished by three characteristics: relatively low-growing basal leaves dominated by unbranched stems bearing uncrowded whorls of smallish flowers, blossoms a varied assortment of colors from yellow to blue and purple, and banner petals with a distinctly ruffled border.

The name of Trevor Kincaid, as in this lupine, has been indelibly linked to the flora and fauna of the Pacific Northwest. My first encounter was in the subject of my Master's thesis, a tiny aquatic fly that bears his name as original taxonomist (*Maruina lanceolata* Kincaid). Born in 1872, Kincaid grew up an avid student of nature, making prodigious collections of plant and animal specimens even before entering the fledgling University of Washington in 1894. Within seven years he was head of the Department of Zoology there and continued as inspired teacher, entomologist, and international science adviser until his death at ninety-seven, having shaped the early development of zoology at the university.

Kincaid's Lupine   1994

Lady Fern *Athyrium filix-femina*

Ferns are distinct from the flowering plants in reproducing by spores rather than by seeds. For this they must have adequate moisture, a requirement lavishly fulfilled in the Pacific Northwest. The region is home to more than forty species of ferns. The large deciduous lady fern is sometimes called swamp fern because of its preferred habitat. The graceful fronds, occurring in clusters, are easy to recognize for their broadly lanceolate outline tapered at both ends. Each leaflet is further incised.

JH: The ferns were a relatively late addition to Bonnie's repertoire, the first ones offered for sale having been printed in 2000. They were planned, in part, with her conservation ethic in mind. She designed them at a size (11" x 14") so that she could mat them with the cut-outs from her larger 16" x 20" prints, avoiding waste. They were also printed as an open edition, meaning that she could have later printed more of the same image, in contrast to the strictly limited and numbered larger prints. Thus the ferns are neither numbered nor dated.

Lady Fern

Large Camas   *Camassia leichtlinii*

The Indian name *quamash* or *camass* persists in the scientific name of this one- to two-foot-tall perennial arising from a deep-seated bulb. Flowers vary anywhere from white to deep blue or violet. *Camassia leichtlinii*, the larger of our two common species, is distinguished by the withering petals twisting together above the seed capsule rather than falling separately. It ranges west of the Cascades from southern British Columbia to southern Oregon and into Sierran California.

The starchy bulb of the camas was a prized staple for native tribes in the Northwest. Care was required to avoid poisonous bulbs of another lily, the so-called death camas (*Zigadenus* spp.). Gathering camas root was the incentive for annual festivities, migrations to harvest grounds where the women dug and prepared bulbs (while the men engaged in sports and games). Handles of the women's digging sticks, fashioned from bone or antler, can be seen in museum collections. The sticks themselves, made of fire-hardened wood crooked and flattened at the end, have not survived so well.

The root was first cooked, either roasted elaborately in covered pits lined with hot stones, or boiled. It was then crushed in mortars and the gummy mass pressed into slabs for keeping. Hungry fur trapper Alexander Henry, in the Willamette Valley in 1814, tells in his diary of trading blue beads with the Kalapuyans for slabs of the nutritious food. As if to remind us of more meager times, each April and May the stately camas colors moist meadows and prairies, roadside ditches, or the vacant lot behind a supermarket, with handsome blue-violet blossoms.

Camas   1993

Licorice Fern *Polypodium glycyrrhiza*

High in trees or on the forest floor, wherever
there is a thick blanket of wet moss, these sparse
little evergreen fronds abound, recognized
by leaflets broadly joined to the midrib in a
zigzag pattern. Chew the rhizome for a taste of
licorice.

Licorice Fern

Maidenhair Fern *Adiantum aleuticum*

This most delicate of ferns is found in the
loveliest of places. Beside waterfalls in shady
humus-rich sites, expect the characteristic
palmate display of fringed leaflets on smooth
black hair-like stalks. Maidenhair is deciduous,
producing new leaves all summer.

Maidenhair Fern

Menzies' Larkspur *Delphinium menziesii*

The Dormouse lay there, and he gazed at the view
Of geraniums (red) and delphiniums (blue),
And he knew there was nothing he wanted instead
Of delphiniums (blue) and geraniums (red).
— A.A. Milne, *When We Were Very Young*

Striving for a mouse's perspective on delphiniums in the flower-carpeted meadow atop Marys Peak one crystal morning in early June, I could appreciate this sentiment. The blue is awesome. Each spring and summer a jubilant succession of wildflowers, otherwise known from such disparate corners as the Willamette Valley floor, the upper Cascade Range, and arid eastern Oregon, adorn this highest peak in the Coast Range. So spectacular is it that the summit has been designated by the U.S. Forest Service as a Scenic Botanical Special Interest Area.

Delphiniums are members of the buttercup family and can be found in the Old World and North America. *Delphinium menziesii* is a widely occurring perennial ranging from British Columbia to southern Oregon on the west side of the Cascade Range, from coastal bluffs to prairies and mountain meadows. Plants tolerate conditions from the very dry to moist, with resulting height variation anywhere from six to twenty-four inches. Flowering is in late May through July.

The scientific name of the genus is from *delphinion,* the Greek word for the larkspur, flowers having the characteristic spur-like nectar tube formed from the upper sepals and petals. The species name honors Archibald Menzies, one of the earliest of the heroic naturalists in the Pacific Northwest. A Scot trained in surgery and medicine, his great love was botany. He made the arduous voyage from England as naturalist and ship's surgeon with Captain George Vancouver to survey the Northwest Coast from 1791 to 1795, chronicling the native people and their languages and collecting botanical specimens—among them this delphinium from "Nova Georgia," the area of Puget Sound.

Larkspur   1995

Monarch Butterfly   *Danaus plexippus*

Aptly named, the beautiful and famous Monarchs are surely the royals among butterflies. Their epic journeys and spectacular gatherings are followed with fascination by a devoted public. Monarchs range over the entire continent, but fall migration southward is channeled into distinct flyways. Eastern and central populations winter in the high fir forests of central Mexico. Most western Monarchs head for California, flying forty miles or more a day. Swarming by the thousands, they come to roost in the Monterey pines of Pacific Grove and the eucalyptus at Santa Cruz and in other colonies from Mendocino to Baja. Around Valentine's Day their offspring and the few remaining migrants begin to move north sporadically, laying eggs on milkweed enroute. From the mid-1800s, these long-distance travelers were found island-hopping across the Pacific all the way to Australia and New Zealand, where colonies persist on introduced milkweeds.

It was a Monarch that helped us acquaint our two young daughters with one of the great wonders of nature—metamorphosis. Containing the chubby and ravenous striped caterpillar on its obligatory milkweed plant, we all watched in awe as it transformed into a static bejeweled chrysalis that subsequently released a splendid adult butterfly. Adult males (top) are distinguished by a patch of scent scales on each hindwing and slightly thinner black outlines of wing veins than the females (bottom).

Brilliant color and large size (up to four-inch wingspan) would make them attractive prey, but the secret of their success lies in the eating habits of the larvae. Feeding on poisonous milkweed, they ingest heart toxins that are transmitted to the body tissues of the adult butterflies, rendering them distasteful and noxious to vertebrate predators. In a classic case of mimicry, the harmless Viceroy butterfly is afforded similar protection simply by resembling the Monarch.

JH: In January of 1998 Bonnie and I were privileged to be on a journey to Mexico guided by Bob Pyle. Had she written this narrative after that trip, it would have expressed our wonder at being in the presence of tens of millions of these magnificent creatures.

44

The Monarch 1995

Moss Tingid *Acalypta saundersi*

This exquisite speck of life keeps company with giants. In the remnant ancient forests of the Pacific Northwest, where three hundred- to one thousand-year-old conifers reach giant proportions, this minute lacebug (here magnified about sixty times) lives in the lush moss that blankets everything on the forest floor. Indeed, the association is so consistent that *Acalypta saundersi* is considered an indicator species, a reliable member of the assemblage of creatures generally associated with a healthy old-growth forest ecosystem.

Bearing witness to a long evolutionary history in a stable environment, this *Acalypta* is flightless—there being no advantage in an ability to fly if one's home has generally been dependable. Many of the vast array of insects associated with old-growth forests have a limited ability for dispersal. In contrast, there are closely related species of moss-feeding lacebugs that are also able to occupy disturbed forest habitats, and some of these individuals are fully winged.

According to the broad classification system devised by Carl Linnaeus in 1758, *A. saundersi* is a true bug (Order Hemiptera) and belongs to the Family Tingidae—an elegant clan in which most of the relatives are cloaked in lace. In the greater scheme of things, it is but one of an estimated eight thousand species of arthropods that occupy the rich mosaic of habitats within the ancient forests of the Pacific Northwest. Who they all are and what each does are mysteries barely explored. Let us learn to appreciate the role of such easily overlooked creatures in what is simply the most magnificent coniferous forest on earth.

Lace Bug   1994

Mountain Lady's Slipper   *Cypripedium montanum*

Slipper orchids world-wide have fascinated scientists and gardeners for centuries, the earliest recorded description dating from 1561. Carl Linnaeus coined the genus name *Cypripedium* after Cyprus, the birthplace of Venus, and *pedilum,* a shoe or slipper. Charles Darwin studied their pollination. And David Douglas, that intrepid Scottish explorer-naturalist sent by the Horticultural Society of London in 1824 to collect exotic flora in the Pacific Northwest, must have been thrilled to come upon an exquisite new variety near the mouth of the Lewis River. He coined the name *montanum* for the species.

This mountain lady's slipper is stunning to behold—two feet tall, of bizarre configuration, retiring in patterns of light filtered through a leafy overstory. The stout stem is clasped by trim lance-shaped leaves pleated longitudinally with prominent veins. One to three distinctive flowers crown the piece. Each is a luminous white pouch watched over by a bright red-spotted yellow staminode and coiffed with long, twisting, maroon-brown sepals and petals. The pouch is a kind of pitfall trap to ensure perpetuation of the species. No nectar is produced. An insect lured into the opening by deceit is detained by the incurved margins and obstructing staminode. Only narrow exit passages are presented, adjacent to the two anthers and the stigma. There the intruder will likely pick up or transmit pollen enroute to freedom.

Prized hardy lady's slippers are now considered rare, their populations having been decimated by overcollecting and habitat destruction. *Cypripedium montanum* is no exception. Historical records list some Oregon sites: "damp roadside near Oswego" in 1887, "hills northwest of Corvallis" in 1907, and "from a school girl" in Lane County in 1939. An innocent bouquet interpreted by present perspectives probably explains why there are none here today. Small isolated populations do persist in moist woods in mountains from southeast Alaska into California and east in Alberta, Montana, and Wyoming. Recognized as vulnerable, *C. montanum* has made some important lists: Oregon Natural Heritage Program "taxa of concern requiring continued monitoring" and U.S. Forest Service and Bureau of Land Management "species to survey and manage." Having hustled this spectacular species along the path to extinction, only we humans can reverse the trend.

Lady's Slipper   1998

Oak Fern *Gymnocarpium dryopteris*

Twice triangular, the entire frond is triangular in
outline and further divided into three triangular
leaflets of delicate incised pinnules. The name is
misleading, as there is no association with oaks.
The deciduous fronds are usually solitary, but
may mass together, carpeting the forest floor.

Oak Fern

Oregon Fawn Lily   *Erythronium oregonum*

My mother calls these "lamb tongues" and recalls, from her childhood in the early 1900s, the thrill of encountering great drifts of them in the moist bottomlands along the Clackamas River. Nearly a century later, April 1994 in Corvallis, I am charmed by isolated plants in Corl's pasture and then by a whole host in Avery Park along the Marys River. Among other picturesque common names for this lily are dogtooth violet, trout lily, Easter lily, adder's tongue, and the current preference, fawn lily.

Erythroniums are native to Europe, Asia, and North America, with the greatest number of species being found in the Pacific Northwest. Indian tribes prized the bulbous root for food, but only along the Pacific Northwest coast was it sufficiently plentiful to be important in the diet. The Kwakiutls of British Columbia cooked and combined it with liberal quantities of fish or whale oil. Raw, it is said to have a milky taste. The name of the genus derives from the Greek *erythro* meaning red, although there are no really red species in the West.

Our species, *Erythronium oregonum,* occurs in moist woods and fields at rather low elevations west of the Cascade Mountains, from southern Oregon north to British Columbia, blossoming from March to May. Arising from corms, the five- to eight-inch stems of *E. oregonum* bear one to three graceful nodding flowers, each with six creamy petals thrust upward. The two basal leaves are strongly mottled, suggesting the fawn lily designation. As it matures, the three-chambered seed capsule is held erect, ultimately releasing a multitude of tiny brown crescent seeds. This Northwest native, with its beautifully patterned leaves and charming flower, is a most satisfactory introduction to home gardens, reproducing rapidly from seed. It favors partial shade with well-drained soil rich in humus.

Fawn Lily   1994

Oregon Iris *Iris tenax*

This showy wild iris, with grass-like leaves and flowers generally in shades of purple, is common in neglected fields and roadsides in the Willamette Valley. As a child growing up near Milwaukie, I used to play in clumps of it, fashioning little lanterns out of the flower parts—and invariably getting a raging case of poison oak in the bargain.

Iris tenax was introduced into the formal world of botany by David Douglas, the Scottish explorer naturalist. Douglas made the voyage to northwest America in 1824, under protection of the Hudson's Bay Company. He found this iris "A common plant in North California, and along the coast of New Georgia, in dry soils or open parts of woods; flowering in April and May." Defined in modern geographic names the range is west of the Cascade crest from Grays Harbor County in Washington State to the Oregon–California border. From material sent by Douglas, John Lindley, Professor of Botany in the University of London, wrote the original description in 1829.

Iris, the Greek word for rainbow, was applied to this genus for its variety of color. The Latin species name *tenax* (tenacious) derives from these observations made by Douglas on uses of the plant:

The native tribes about Aguilar [Umpqua] river, in California [Oregon], find this plant very serviceable for many purposes: from the veins of the leaves fine cord is made, which is converted into fishing nets; and from its buoyancy, great strength, and durability, it suits this purpose admirably. It is also made into snares for deer and bears; and a good idea may be formed of its strength when a snare, not thicker than a 16-thread line, is sufficient to strangle *Cervus Alces* [elk], the Great Stag of California, one of the most powerful animals of its tribe. The cordage is also manufactured into bags and other articles.

Flags   1992

Oregon Iris II   *Iris tenax*

JH: Flags (1992), reproduced on the previous
page, was the first print that Bonnie made.
That, combined with technical difficulties that
resulted in only sixty-five acceptable prints from
a run of one hundred, caused this print to sell
out early. She then decided to do a new edition,
a lighter purple Flags II, demonstrating some of
the color variation in the species and her point
that "somewhere out there is a flower just this
color."

Flags II   1998

Oregon Silverspot Butterfly   *Speyeria zerene hippolyta*

Here is the only butterfly in Oregon to be listed as a threatened species under the Endangered Species Act. Its habitat is in those natural salt spray meadows along the Oregon coast north of the Florence dunes where its preferred larval food plant, fragrant little *Viola adunca*, is found. As construction, overgrazing, or encroachment of brush wipe out the violets, the butterflies too are lost. Only two stable examples of this ecosystem supporting *Speyeria zerene hippolyta* remain.

The beautiful orange-brown fritillaries of the genus *Speyeria* are fascinating for their geographic variability. The *zerene* species alone has evolved six recognized subspecies occurring in Oregon. One of these is our *S. z. hippolyta*, unusual in its adaptation to rigorous windswept coastal meadows. Surely it is this combination of hardiness and beauty that inspired naming it after Hippolyta, the Queen of the Amazons in Greek mythology. All members of the genus *Speyeria* feed on various kinds of violets.

Eggs of *S. z. hippolyta* are laid in August and September on or near the food plant. Newly hatched larvae spend the winter in hibernation and commence feeding the following spring. The dark, mottled, spiny caterpillars are mainly nocturnal and seldom seen. Pupation occurs after five molts and adults are late flyers, not appearing in their flower-filled meadows until August and September. Both sexes of this elegant two-inch fritillary have the orange background and dark checkered wing pattern above, but on the underside of the hindwings there is treasure— the spots of silver.

Endangered Silverspot   1993

Oregon Swallowtail Butterfly   *Papilio oregonius*

This strikingly beautiful swallowtail is a true native of the Northwest. It was first described in 1876 from a specimen collected near The Dalles by Henry Edwards, an enthusiastic lepidopterist, trustee and vice president of the San Francisco Academy of Sciences, as well as an eminent actor on the San Francisco stage. For description and naming, he loaned the specimen to William H. Edwards (no relative), the foremost student of American butterflies.

The species can be found east of the Cascade mountains in the lower basalt and sagebrush canyons, slopes, and hilltops of the Columbia and Snake river basins in Oregon, Washington, and southern British Columbia. The larvae feed on tarragon sage (*Artemesia dracunculus*), an unusual choice because its closest relatives feed on members of the parsley family (Umbelliferae). The adults nectar from many flowers, especially thistles. The males may be found at mudpuddles, and coursing down hot canyons or hilltopping in search of the females.

These butterflies are wary and strong fliers. The species can be recognized by the characteristic swallowtail shape and distinguished from the similar Anise Swallowtail (*P. zelicaon*) by its larger size and color of the abdomen—*P. oregonius* has a yellow abdomen with black stripes, *P. zelicaon* a mostly black abdomen with a yellow stripe down each side.

Recognizing aesthetic, educational, and regional values, the Legislative Assembly of the State of Oregon in its 1979 regular session adopted by Senate Concurrent Resolution the Oregon Swallowtail butterfly (*Papilio oregonius*) as Oregon's official state insect. The U.S. Postal Service had brought this splendid butterfly to the attention of the American public through the issuance of a postage stamp two years earlier, one of a set of four illustrating selected American butterflies. This exposure considerably aided the advocates' effort to convince the legislators.

Oregon Swallowtail   1989

Spring Beauty, Slender Toothwort   *Cardamine pulcherrima*

Botanists wisely use the scientific names of plant species, thereby avoiding much confusion. The rest of us are prone to calling our wildflowers by some common, often regional, name. "Spring beauty" is a natural choice for the first flower to bloom after winter. In the high Cascades quite a different plant, blooming close on the heels of melting snow, claims this common name. But in the woods of western Oregon, *Cardamine pulcherrima* is our spring beauty.

This first of blossoms can be found from early February in moist open woods, mostly west of the Cascade crest from British Columbia to northern California. Color varies from lavender to pink, or even white. Stems occur singly and range from four to eight inches in height, arising from a short rhizome. The round basal leaves are seemingly unrelated to the flower stems.

As if to prove that scientific names too can be fallible, this plant was originally designated as *Dentaria tenella* when described by Frederick T. Pursh in 1814. For twelve years this German botanist had traveled and lived in America as collector, gardener, and landscape architect.

C. pulcherrima is a member of the mustard family, Cruciferae or cross bearers, so named for their characteristic four petals as likened to the four arms of a cross. Thus this little harbinger may be thought of as each spring bearing its delicate cross at the head of the jubilant procession of wildflowers to follow.

JH: Further adding to the naming dilemma, this species is now universally called *Cardamine nuttallii.*

Spring Beauties   1992

Stream Violet, Johnny-Jump-Up   *Viola glabella*

This bright perennial belongs to the large and widely distributed family Violaceae (violets). A familiar garden relative is the pansy, originally cultivated from a wild European violet. In the Pacific Northwest, violet species have adapted to virtually every major habitat type, from wet to dry and sea level to subalpine. *Viola glabella* is common from sea level to 8,000 feet in moist woods and along streams, from Alaska to central California, east into Idaho and Montana, and in Japan. It was Thomas Nuttall who first collected this species "in shady woods of the Oregon [Columbia River]" and described and named it for science in 1838. Nuttall, an English explorer botanist famed for his prodigious list of discoveries and descriptions, had resigned his post at Harvard to make the arduous trip west with Nathaniel Wyeth's 1834 expedition.

The name *Viola* (an ancient Latin name for the violet plant) and *glabella* (from "glabrous," meaning smooth) has prevailed unaltered. This is a stemmed violet, leaves and flowers branching from a common stem, three to twelve inches tall. Of the five bright lemon-yellow petals, the two lateral and the lower are finely marked with dark purple "bee lines." The lower petal is drawn out behind into a short spur or nectar sac. A cluster of short hairs (beard) at the base of the lateral petals affords a foothold for pollinating insects. Both basal and stem leaves are generally hairless, heart-shaped, and edged with rounded indentations. The oblong seed capsule contains three ranks of tiny brown seeds that are ejected forcibly when mature.

I learned to call these johnny-jump-ups, perhaps because of their sudden appearance in winter-weary woods. Along with spring beauties and trilliums, they liven the forest floor before trees and shrubs have leafed out. These violets transfer well to the home garden where, given adequate encouragement, they may even become invasive. As if to compensate for lacking fragrance, they sport the color of sunshine and a jaunty growth form. I can't help wondering how we would define the color "violet" if it had been inspired by such a yellow violet rather than a purple one.

Stream Violet   1997

Sword Fern *Polystichum munitum*

Here is our quintessential fern of the deep forest—tall, tough evergreen fronds in big circular clusters. At the base of each serrated sword-like leaflet is a distinctive forward-pointing lobe. Rusty brown clusters of spores dot the undersides.

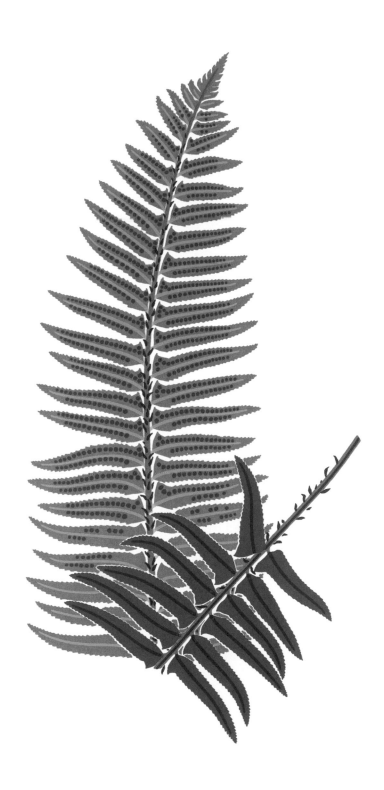

Sword Fern

Tiger Lily, Columbia Lily, Oregon Lily *Lilium columbianum*

The most spectacular company of tigers I know grows in the summit meadow on Marys Peak. There a particularly sturdy ecotype has developed in response to the rigorous environment. In spite of heavy foraging by marauding deer with voracious appetites for lily buds, the meadow in late June and early July can be a sea of tiger orange. Originally described for science from plants collected by David Douglas near the Columbia and Willamette rivers, *Lilium columbianum* ranges from southern British Columbia to northern California and east to Idaho and Nevada in open damp woods and meadows, from sea level to 6,000 feet.

Tiger lily is a common name shared by a number of species bearing the familiar bright orange petals and sepals, recurved (like a Turk's cap) and freckled with purple spots. The distinguishing characteristics of *L. columbianum* are the relatively small blossoms (one to two inches) and stamens parallel rather than flared. The stiff one- to four-foot stem, arising from a deep-seated white bulb, is interspersed with narrowly lance-shaped leaves mostly in whorls, and is crowned with two to twenty showy blossoms hanging bell-like on long pedicels. Nectar, secreted at the base of the ovary, fuels butterflies and hummingbirds. The fruit is a three-chambered oblong capsule containing numerous flat triangular seeds.

Bitter, peppery tiger lily bulbs were steamed and dried by native tribes and used as a condiment in stews. Contemporary use is restricted to the aesthetic. Consider adding this radiant perennial to a home garden. Since the bulbs do not transplant well and digging them is not appropriate, native plant nurseries are the best source. You can grow plants from seed, given sufficient patience to wait three to five years for flowering. Or simply revel in tigers in the wild.

Tigers   2000

Tricolored Monkeyflower   *Mimulus tricolor*

A low-lying plant less than five inches high made front-page news in the *Corvallis Gazette-Times* in May of 1999 as "A Treasure from the Past." Even the prestigious journal *Science* proclaimed " 'Extinct' Oregon Flower Reappears." Once so common on the lowland prairies of the Willamette Valley that masses of the tiny trumpet-shaped flower blanketed fields in purple each spring, *Mimulus tricolor* had not been seen here since 1991. It reappeared, to joyous acclaim, in a former rye grass field after winter flooding of the Marys River had stripped away the sod and left the requisite vernal pools.

There are many species of *Mimulus* in western North America. The genus is also native to Asia, Africa, Chile, Australia, and New Zealand. I always expect to find monkeyflowers in exquisite wet places, their bright faces clustered beside crystal mountain streams or in mossy seeps. They come in various colors—yellows, purples, and even red. The generic name is a diminutive of the Latin *mimus*, a comic actor.

M. tricolor may be tuned to very specific requirements of wetness. It is associated with early spring ponding of water where growth, flowering, and seed set must be accomplished rapidly before the clay substrate dries, cracks, and becomes inhospitable. Seeds of this species are uniquely adapted to survive dry periods. Encased in a hard nutlike capsule, they can remain dormant until released by flooding. Stream engineering, agriculture, and development have largely eliminated suitable habitat. Isolated microcosms with favorable conditions prompted a reprise. Old seeds, stored in the seed bank in the soil, germinated. Clusters of delicate purple flowers ensued, captivating the community. Was this just a fleeting glimpse of past splendor, or are the little tricolored monkeyflowers back to stay?

Monkeyflower   1999

Twinflower *Linnaea borealis*

This tiny treasure, with its nodding twin flowers, is found in cool moist woods in boreal regions the world around. It is the namesake and signature flower of Carl Linnaeus, Swedish botanist, physician, and teacher, who brought order to natural science by devising a uniform system for the classification of plants and animals. Linnaeus asssigned two-word Latin names to some 7,700 plant species. His work is the starting point for all modern botanical nomenclature. *Linnaea borealis* is therefore of special significance to botanists everywhere.

Linnaeus first encountered twinflower while exploring the wilds of Lapland in 1732. He was so attracted to this modest plant that it became the centerpiece of his family crest, and he is typically pictured with a sprig of the little flower in his hand. He described it as of short growth, easily overlooked, flowering for only a brief period. When it was named after him he wrote, "it is called after Linnaeus, who resembles it." It is appropriate that a Swedish medal struck to commemorate the 250th anniversary of his birth features *L. borealis*.

In the Northwest, twinflower may be discovered in such places as along a mountain trail in the Cascades or a wooded roadside on an island in Puget Sound. Its creeping stems, woody at the base, carpet the forest floor with glossy evergreen leaves. Numerous erect leafy stems of less than four inches terminate in pairs of delicate bell-shaped pink flowers, one-half-inch long, borne on slender stalks. The fruit is roundish and contains one seed. The leathery leaves are opposite, oval, shallow-toothed, and conspicuously veined. Look for flowers from June to September.

Twinflower   1992

Western Blue Flag, Rocky Mountain Iris   *Iris missouriensis*

In his comprehensive field guide to the *Wildflowers of the Columbia Gorge*, photographer Russ Jolley says this iris is to be found at the trailhead to Horsethief Butte, flowering in mid-May. So we stopped right beside the highway that follows closely the north side of the great river, and there it was, precisely as predicted. It is a tall and stately creature rooted in vernal seeps in an otherwise arid landscape.

This is the niche it occupies throughout its range east of the Cascade crest in sagebrush desert and pine forest from British Columbia to Southern California and east to the Dakotas, always where there is a source of moisture until flowering. So dependable is it as an indicator of ground water close to the surface that ranchers in d mountain country chose to dig watering holes for livestock in places where it occurred. Native tribes also enlisted the plant, but for a more hostile purpose. Arrowheads were dipped in a concoction made of the ground root and animal bile. It was reported that warriors only slightly wounded by these arrows soon died.

A confusion of wild irises exists along the Pacific Coast, but east of the Cascade Mountains *Iris missouriensis* is the one common and unmistakable iris species. It was described and named in 1834 from specimens collected "toward the sources of the Missouri." An intriguing puzzle remains. On Whidbey and the San Juan islands in Washington, far from its defined range, sporadic outlier populations of *I. missouriensis* have been found. Can they be hangers-on from a former time (5,000 to 8,000 years ago) when the maritime climate was warmer and drier? Having persisted for millennia, these rare little colonies are now dwindling as their island habitat succumbs to development.

Western Blue Flag   1997

Western Trillium, Wake-robin   *Trillium ovatum*

While other vegetation is still winter-bare, this elegant white blossom unfurls in its whorl of large sessile leaves to punctuate moist shady woods from British Columbia south to central California, from lowlands to well up in the mountains. Indeed, one of its common names, wake-robin, implies that it precedes the earliest of birds. The Latin generic name *Trillium*, meaning triple, is appropriate to describe the parts of three in petals, sepals, and leaves—all supported by a sturdy bare stem as much as twelve inches tall. This species is distinguished from another common woodland trillium in having a thin stem or pedicel that bears each solitary flower above its three broad net-veined leaves. The fragrant white flowers of *Trillium ovatum* turn purplish with age.

The botanical history of this plant is tied to the history of the Pacific Northwest via the Lewis and Clark Expedition. On their return trip east in 1806, Captain Meriwether Lewis collected it "On the rapids of Columbia river," one of 150 novel specimens he conveyed to botanist Frederick Pursh. It was from the Lewis collection that *T. ovatum* was first described and named by Pursh in his *Flora Americae Septentrionalis* of 1814, the first account of North American plants to include the Pacific Northwest.

The familiar admonition not to pick trilliums is well founded. Removing the flower stem robs the rhizome of the food supply necessary to produce the next year's plant, and some years may be required for recovery. Trilliums do not transplant well, but may be grown from seed with patience. As many as seven years may pass before a white blossom is produced to crown the three stem-leaves. Ants are a natural aid in dispersal, attracted by sweet tissue on the seeds.

Trilliums   1993